MCR

MCR

Science Investigations

SOLIDS:
AN INVESTIGATION

CHRIS OXLADE

PowerKiDS press.

New York

Published in 2008 by The Rosen Publishing Group, Inc.
29 East 21st Street, New York, NY 10010

First Edition

The publishers would like to thank the following for permission to reproduce these photographs:
Corbis: 4 (Doug Wilson), 6 (Steve Kaufman), 18 (Lawson Wood), 22 (Otto Rogge), 24 (Jacqui Hurst), 26 (Layne Kennedy), 28 (Hubert Stadler); OSF/Photolibrary: 5 (Workbook, Inc.), 7 (Foodpix), 8 (Maximilian Stock Ltd), 9 (Index Stock Imagery), 10 (Phototake Inc.), 11 (Bildhuset Ab), 12 (Jon Arnold Images), 13 (Norbert Rosing), 16 (Doug Allan), 17 (Index Stock Imagery), 19, 20 top (Index Stock Imagery), Cover and 20 bottom, 23 (Australia), 25 (Workbook, Inc.), 27 (Index Stock Imagery), 29 (Stock4b Gmbh); Redfern: 14; REX Features: 15 (SIPA).

Editors: Sarah Doughty and Rachel Minay
Series design: Derek Lee
Book design: Malcolm Walker
Illustrator: Peter Bull
Text consultant: Dr. Mike Goldsmith

Library of Congress Cataloging-in-Publication Data

Oxlade, Chris.
  Solids : an investigation / Chris Oxlade. — 1st ed.
     p. cm. — (Science investigations)
  Includes bibliographical references and index.
  ISBN-13: 978-1-4042-4284-5 (library binding)
 1. Solid state physics—Juvenile literature. 2. Matter—
Properties—Juvenile literature. 3. Solids—Juvenile literature. I.
Title.
  QC176.3.O95 2008
  530.4'1—dc22

                                        2007032606

Manufactured in China

Contents

What are materials?

In science, the word *material* means "a substance." All the things in our world, from plants and animals to parts of computers and aircraft, are made from some kind of material. There are literally millions of different materials. Each material has different properties from all the others. They may be hard and strong like steel, or soft and flexible, like rubber. Steel and rubber are solid materials. Materials can also be liquids and gases.

Wood is a natural material. Softwood, like these logs, comes from conifers. It is easy to cut and shape, strong and easily renewable.

Some materials are natural or "raw," such as rocks, clay, wood, wool, and silk. Others are made from raw materials and include metals extracted from ores, and plastics made from oil and natural gas. Many materials have similar properties to others, so we can classify them as a group. Metals, for example, tend to be shiny and heavy, and are good conductors of electricity and heat. Other groups of materials include rock, wood, and plastics.

INVESTIGATION

What materials are used in your home?

MATERIALS

A selection of objects made from a variety of different materials (such as metal, plastic, paper, and wood).

INSTRUCTIONS

Gather up a selection of objects made from different materials.

Look at the materials the objects are made from. Try to make groups of objects that you think are made from similar materials. For example, sort the objects into metals, plastics, paper, and wood.

Write down the properties of the material in each group you have made. For example, whether the objects are shiny, heavy, light, weak, or strong.

FURTHER INVESTIGATION

Look at each object you have listed and try to decide why it is made from its material. What properties of the material made it suitable for making the object? Could the object have been made from a different material?

These plastic bowls and plates are light and strong. Plastic is a synthetic material made in factories—but it is made from oil and natural gas, which are found naturally in the ground.

5

What are "states of matter"?

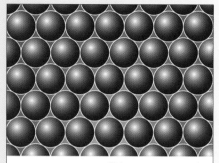

solid

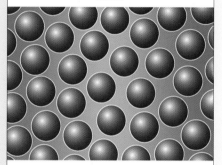

liquid

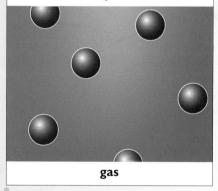

gas

Solids, liquids, and gases are all made of tiny particles. This diagram shows how the particles are arranged in each state.

Look at the picture. Can you see water in all its three states?

EVIDENCE

Materials exist in one of three different forms. These forms are solids, liquids, and gases. For example, wood is a solid, oil is a liquid, and air is a gas. The words *solid, liquid* or *gas* describe a material's "state of matter." Everything on Earth is in one of these three states. However, many materials exist in all three states under different conditions. We usually see water as a liquid that we drink, or see it flowing down a river bed or in a lake or pond. But water also exists as a solid (when it is called *ice*) and a gas (when it is called *water vapor*).

We recognize solids, liquids, and gases by their properties. Solids are materials that do not usually change their shape or volume (although they can be soft, like margarine). Liquids are materials that flow but keep their volume. This is why liquids always fill the bottom of any container they are in. Gases do not maintain their shape or their volume. They are materials that flow freely and expand to fill any space they are in.

INVESTIGATION

How well do solids, liquids, and gases flow?

MATERIALS

A pan (either a glass pan or a metal pan with a glass lid) and ice cubes.

INSTRUCTIONS

Put a few ice cubes in the bottom of a pan and put the lid on. Ice is a solid, so the cubes keep their shape and do not flow.

Heat the pan gently so that the ice melts. Water is a liquid, so it flows to fill the bottom of the pan. It fills the same space, or volume, as the ice. Keep heating it until the water boils.

Turn off the heat when almost all the water has gone. Water vapor is a gas so it flows and expands to fill the pan. It has a much bigger volume than the ice or the liquid water.

WARNING: Be careful with boiling water. Get an adult to help you.

How many states of matter does this carbonated drink contain?

FURTHER INVESTIGATION

In some machines, the pressure of liquid is used to move the machine's parts. The liquid is forced along a tube so that it moves a piston in or out of a cylinder. Can you find out what name is given to machines that work in this way?

A road drill is a pneumatic machine. It uses a material in a compressed state to provide power for it to work. What is this material?

Which materials are strongest and hardest?

Strength and hardness are important properties of many materials that are solid. Very strong materials are used to make machines such as aircraft and structures such as skyscrapers. Very hard materials are used to make cutting tools, such as knives and saws, and long-lasting objects, such as kitchen worktops and floors.

Car bodies are usually made of steel. Steel is an alloy, made up of several materials—mainly iron. Iron by itself is quite soft and weak, but adding other ingredients such as carbon makes it hard and strong.

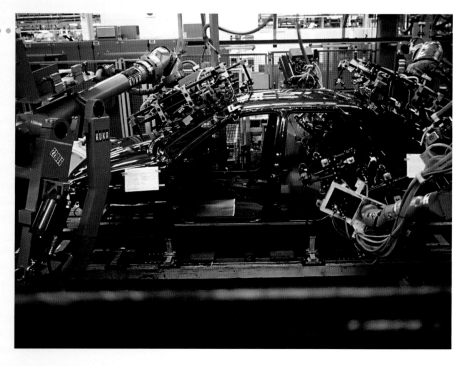

All materials are made up of tiny particles (see page 6). Each particle in a solid is joined to the particles around it. This is why solids keep their shape and do not flow like liquids or gases. The more strongly the particles are joined together, the stronger and harder the material is. The particles in wood, stone, and many metals are strongly joined, and this makes them difficult to bend or break. In weak, soft materials, such as drawing chalk, the particles are weakly joined. When a solid melts, the bonds between the particles break.

INVESTIGATION

Which materials are hardest and which are softest?

MATERIALS

A selection of objects made from different materials (such as metal, soap, wood, plastic, card, and stone). Only use old objects that will not be spoiled if they are scratched.

INSTRUCTIONS

Choose two objects made from different materials. Try to scratch a mark on one object with the other. This is called a *scratch test*. The object that is not marked is made of the harder material.

Keep the object of the harder material and test it against another object. Keep testing, and each time keep the harder material. Eventually, you will have worked out which material is hardest of all.

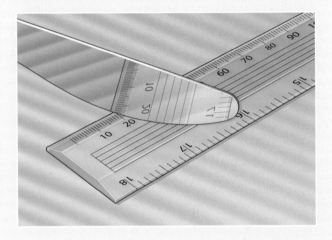

Glass is a brittle material, which means it is hard but also breaks easily. Sand, limestone, and soda ash are mixed and melted together at high temperatures to form solid glass.

FURTHER INVESTIGATION

Using scratch tests, arrange your objects in order of the hardness of the material they are made from. Make a list of the materials with the hardest at the top.

A diamond is one of the hardest materials there is. What is the only material that can be used to cut a diamond?

How does electricity get through materials?

This is the French TGV—
a train that can reach
very high speeds. It runs
on a special track and is
powered by overhead
power lines.

Some materials let electric current flow through them. They are called *electrical conductors*. Metals are the best electrical conductors. We make use of them in electric circuits and electronic circuits. The copper wire inside an electrical cable is a conductor that carries electric current along the cable.

However, most materials do not conduct electricity at all. They are called *insulators*. Even though insulators cannot carry electric current, they are useful in electric circuits. Electrical cables are covered in plastic (an insulator) in order to stop the electricity jumping accidentally to other cables.

Water is a poor conductor of electricity, but it is not an insulator. A certain amount of electricity can flow through it. This is why it is dangerous to touch light switches or mains-powered machines with wet hands. Electricity can also flow through you, so you need to be careful.

Can you see where the ceramic insulators are on electricity pylons? They look like a stack of large plates. They stop the electricity from flowing out of the cable, into the pylon, and down to the ground.

INVESTIGATION

Which materials conduct electricity?

MATERIALS

A selection of objects made from different materials, a 1.5 V battery, connecting wires, a bulb, and a bulb holder.

INSTRUCTIONS

Make up the circuit by connecting the bulb, battery, and wires. The bulb should light up if the circuit is completed.

Choose an object and touch the ends of the loose wires on different parts of the object. If the light bulb glows, it means that the object has completed the circuit and an electric current is flowing. This means the object is a conductor. If the light does not glow, it means the object is an insulator.

Make a record of your findings.

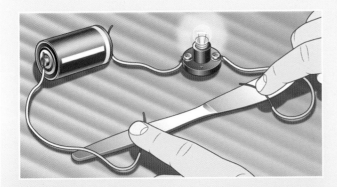

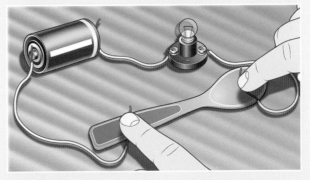

FURTHER INVESTIGATION

Group the objects into conductors and insulators. Which family of materials are good conductors?

Can you name a material that is not a metal but a good conductor? (Clue: it is also found in pencils.)

How does heat get through materials?

When you stir a hot drink with a metal spoon, you can feel the spoon handle getting warm. This is because heat is moving along the spoon from the drink and into your hand. Because the heat can travel through the spoon, the metal that the spoon is made of is called a *good conductor of heat*. Metals are good conductors of heat. We use good conductors of heat to help heat get from one place to another. For example, radiators are made of metal, so that the heat from the water inside can get out into the room. Other materials do not let heat pass through them at all. They are called *good insulators*. If one part of an insulator is heated up, the other parts stay cool. Liquids and gases are both good insulators.

The roofs of these houses are covered with turf. The turf is a good insulator. It helps to stop heat escaping from inside the house.

We use insulators to stop heat escaping from our bodies. When it is cold, we wrap thick clothes around ourselves. This traps a layer of air that helps to keep us warm. Animals use their thick coats to survive cold conditions. Fluffing their hair or fur traps a layer of air around their bodies, keeping them warm.

Which materials are good heat conductors?

MATERIALS

A mug, small plastic beads (or similar objects), butter, a metal spoon, a plastic spoon, a paper straw, a plastic straw, a glass rod, a wooden rod, and a styrofoam rod. The objects you are going to test all need to be the same length.

INSTRUCTIONS

Put a small blob of butter on the upper end of each of the items. Place a bead on the butter so that it sticks.

Ask an adult to help you pour boiling water into a mug. Stand the objects in the mug with the beads at their tops.

Watch what happens. The heat energy will travel along the objects that are good conductors, making the butter melt and the bead will then fall.

The first bead to fall will be from the best conductor. Put your objects in order of the best to worst conductor.

Animals, such as this Arctic fox, and birds are adapted to keeping out the cold. Tiny muscles in the skin pull their hair, fur, or feathers upright to make an insulating layer.

FURTHER INVESTIGATION

In cold countries insulating materials are used to stop valuable heat escaping from houses and homes. How many examples of insulating materials can you find in your home? Make a list.

What happens when materials are heated?

We often see materials change from one state of matter to another. For example, water changes to steam in a kettle. This change is called a "change of state." Changes of state normally happen when the temperature of a material changes.

If you keep heating a solid, its temperature keeps going up. At a certain temperature, the solid usually turns into a liquid. This change of state is called *melting*, and the temperature at which it happens is called the material's *melting point*. If you keep heating a liquid, its temperature keeps going up. At a certain temperature the liquid turns into a gas. This change of state is called *boiling*. The temperature at which it happens is called the material's *boiling point*. Different materials have different melting and boiling points. For instance, the melting point of ice is 32°F (0°C), but the melting point of iron is a massive 2,804°F (1,540°C). Some substances, such as carbon dioxide, do not have a liquid state at all. Instead they turn from a solid into a gas, without first melting. This is called *sublimation*.

At this concert, solid carbon dioxide (called *dry ice*) sublimates to form carbon dioxide gas.

INVESTIGATION

What is the boiling point of water?

MATERIALS

A pan, a cooking thermometer (which should measure beyond 212°F or 100°C), and a wooden spoon.

INSTRUCTIONS

Half fill a pan with warm water. Put the pan on a stove and heat it gently. Keep stirring the water with the wooden spoon and keep measuring the temperature with the thermometer.

The temperature will gradually rise. Eventually, it will stop rising and the water will boil.

Write down the temperature at which this happens.

FURTHER INVESTIGATION

Try the experiment again. Measure the temperature every minute and write it down. Keep boiling the water gently. Does the temperature rise above 212°F (100°C)? Draw a graph of temperature against time.

Do you think that the number of particles that make up the liquid changes when water turns into a gas?

If this ice cream is made mostly of water, what do you think its melting point would be?

What happens when materials are cooled?

Changes of state also happen when the temperature of a material goes down. If you cool a gas, its temperature gradually falls. At a certain temperature the gas turns into a liquid. This change of state is called *condensation*. It can happen during cooking, when steam condenses on the cooler lid of the saucepan. The temperature at which this happens is the same as the material's boiling point. Steam condenses into water in the pan at 212°F (100°C).

Materials that are normally gases, such as air, often have to be cooled to a very low temperature before they turn to liquid. Oxygen, one of the gases in the air, boils (and condenses) at -360°F (-218°C). If you keep taking away heat from a liquid, its temperature keeps going down. At a certain temperature the liquid turns into a solid. This change of state is called *freezing*. The temperature at which this happens is the same as the material's melting point. For pure water, under normal conditions, the freezing point is 32°F (0°C), the same as the melting point of ice.

. . . .

This picture shows an iceberg in the Canadian Arctic. Sea water freezes at a slightly lower temperature than fresh water because of the salt in it.

What happens to air when it is cooled?

MATERIALS

A freezer, a party balloon, and tape.

INSTRUCTIONS

Blow up a party balloon and tie the neck to stop the air from leaking out. Carefully stick a piece of tape around the widest part of the balloon.

Put the balloon in a freezer. After a few minutes look at the tape. It will have crumpled, showing that the volume of the air has gone down.

Remove the balloon from the freezer and watch what happens as the air inside is warmed up again.

Does air expand or contract when it is cooled?

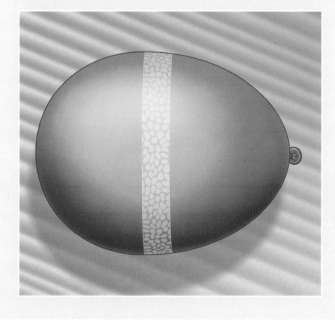

Heating the air inside a balloon makes the air expand, and the balloon rises. As the air cools again, it shrinks and the balloon descends.

FURTHER INVESTIGATION

What do you think would happen if you put the balloon in a bowl of warm water?

Can you find out another way of making the volume of a gas change, apart from changing its temperature? (Clue: think about a bicycle pump.)

What happens when a solid dissolves?

When you put sugar in a drink and stir it, the sugar crystals seem to gradually disappear. This process is called *dissolving*. We say that the sugar has dissolved in the liquid. When a material dissolves in a liquid, the tiny particles that it is made of break off and get mixed with the particles of the liquid that it dissolves in.

The mixture of particles made when a material dissolves in a liquid is called a *solution*. The material that dissolves is called the *solute* and the liquid is called the *solvent*. Because sugar dissolves in water, we say that sugar is soluble in water. After a certain amount of sugar has been dissolved, the solution becomes saturated and no more material will dissolve in it.

Many materials, such as plastics and paints, do not dissolve in water. They are said to be *insoluble*. But they will dissolve in other types of solvent. For example, paint stripper contains a solvent that dissolves paint, allowing it to be scraped off.

Water does not dissolve oily substances. To wash them off, we add oil-dissolving chemicals called *detergents* to the water. This tractor is spreading chemicals at the site of an oil spill.

INVESTIGATION

What happens when you mix salt with water?

MATERIALS

A glass, salt, and a spoon.

INSTRUCTIONS

Half fill a glass with water. Add several spoonfuls of salt and let it settle. See how high it reaches up the sides of the glass. Then give it a good stir and, when it's settled again, see how far the height of the salt has fallen.

Some of the salt has dissolved to make a salt solution. The salt seems to have disappeared, but its tiny particles are mixed in with the water particles. The remainder of the salt has not dissolved, because the water has reached a point where it cannot hold any more salt. It is now a saturated solution.

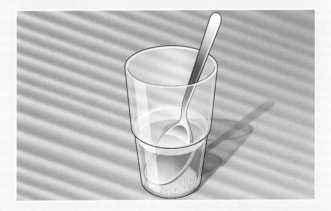

Bubbles appear when a carbonated drink is opened. Where does the gas come from?

FURTHER INVESTIGATION

Try the experiment again, this time using very hot water from a kettle. Does the salt dissolve more quickly than it did in the cold water?

Do you think salt will dissolve in cooking oil? Try it and see.

Can changes in materials be reversed?

This picture shows the world's largest geode, a hollow rock with crystals lining the inside walls. These crystals are made of a mineral, which was dissolved in water. The water flowed into the cave and the mineral turned back to a solid.

You can make ice cubes for cold drinks by putting water in a freezer to cool them below 32°F (0°C). When you put the ice cubes in a drink, they gradually turn back into water. So water can be turned from liquid to solid and back again. The same is true of water when it boils to make water vapor. The water vapor can be turned back to water if it is cooled.

These physical changes are all reversible. The change of state caused by heating or cooling a material can be changed back. This is done by reversing the change in temperature. Dissolving is also usually a reversible change. If you dissolve sugar (the solute) in water (the solvent), the sugar disappears. However, heating the water can cause the water to evaporate and the sugar crystals can be recovered. In other words, the solute can be returned to its earlier state if the solvent is removed from the solution.

Metal objects are made by pouring red-hot, liquid metal into molds, as shown here. When they cool, they take on the shape of the mold. This is called *casting*.

INVESTIGATION

How can you reverse condensation?

MATERIALS

A tall glass jar, a bowl, kitchen foil, and ice.

INSTRUCTIONS

Pour an inch or two of water into a tall glass jar. Cover the jar with kitchen foil and make a dip in its center. Put a few ice cubes in the dip. Stand the jar in a

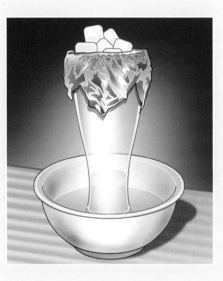

bowl of hot water and watch what happens in the jar.

Water evaporates from the bottom of the jar to make water vapor. When the water vapor hits the cold foil, the evaporation is reversed. Water is formed again by condensation.

FURTHER INVESTIGATION

The water cycle is the circulation of water along rivers, in and out of seas and oceans, and through the atmosphere. Look at this diagram of the water cycle. Can you identify the reversible changes?

rain

clouds form

evaporation

vegetation

water runs off land

river

sea

lake

What are chemical changes?

During a physical change (such as melting, boiling, or dissolving), a material changes shape or volume. However, its chemical makeup stays the same. But if a piece of wood burns, it changes into ash. This is a completely new material. Burning is an example of a chemical change. Chemical changes are also called *chemical reactions*.

Many everyday changes are chemical changes. Food goes through a chemical change when it is no longer fresh. Fallen leaves go through a chemical change when they begin to rot.

Chemical changes are used in factories to make synthetic materials, such as plastics, from natural materials such as oil. Synthetic materials are designed to have better properties (such as strength, density, and flexibility) than natural materials.

The chemicals needed to make a firework display are mixed together and stored in the firework. A fast chemical reaction takes place in the air when the firework is lit.

INVESTIGATION

What happens when you mix vinegar and baking powder?

MATERIALS

A party balloon, a small plastic bottle, a funnel, baking powder, and vinegar.

INSTRUCTIONS

Push a funnel into the neck of a party balloon.

Pour two tablespoons of baking powder into the funnel so that it falls into the balloon. Pour a dessertspoonful of vinegar into a plastic

bottle. Stretch the neck of the balloon over the neck of the bottle. Now lift the balloon to make the powder fall into the vinegar.

The powder and vinegar fizz, making gas that inflates the balloon. This is evidence that a chemical reaction is happening.

FURTHER INVESTIGATION

The balloon gradually blows up as the reaction happens. Where do you think the gas that blows up the balloon comes from?

A foam fire extinguisher spurts out gas-filled foam. Can you guess what makes the gas?

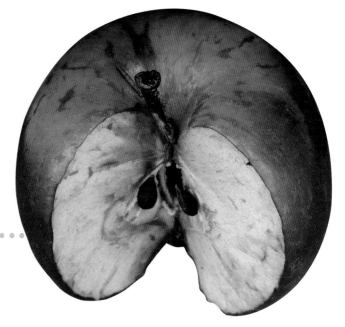

An apple slowly turns brown after it is cut. This is because of a chemical reaction between the air and chemicals in the apple.

Are some changes permanent?

When an egg cooks, the egg white changes from being runny and transparent to being solid and white. The material in the white has gone through a chemical change. If the cooked egg is cooled, the white does not change back to being runny and transparent. The change it has gone through is an irreversible, or permanent, change. The original material of the egg can never be recovered.

The raw batter for a cake is going into the oven. What will happen to the batter when it is heated? Can it ever be changed back?

Most chemical changes are irreversible changes. Rusting is an irreversible change. The iron that turns into the brown, flaky rust can never be changed back. The powdery plaster of Paris used to make casts for broken limbs can never be returned to, once water has been added to it to make it set hard. Some physical changes are irreversible, too. For example, if you saw up a piece of wood into sawdust, the sawdust can never be returned to its original form.

INVESTIGATION

What happens to candle wax when a candle burns?

MATERIALS

A small candle, glass jar, modeling clay, and an old plate.

INSTRUCTIONS

Stand the candle in the middle of the old plate. Put a strip of clay around the rim of the jar (this will make an airtight seal with the plate). Light the candle, put the jar over the top, and press it down so that the modeling clay presses onto the plate.

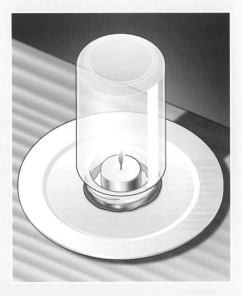

The candle will burn for a few seconds and then go out. Lift the jar. Some of the candle wax has gone and there is some water on the inside of the jar. The candle wax has changed to water and a gas called carbon dioxide, which you cannot see.

Why do you think the candle went out after a few seconds? What gas has been burned up?

FURTHER INVESTIGATION

What three things are needed for burning to take place? Think about lighting a candle and what happens in the investigation above.

How many changes of state can you identify in a candle?

When a candle burns, some wax disappears, but some wax goes through a reversible change. Can you see where this has happened?

How do we separate mixtures?

It is often useful to be able to separate the different materials in a mixture from each other. Peas and the water they are cooked in are a mixture. You separate this mixture by pouring it through a colander. A similar method is used to get large stones out of soil. The mixture of stones and soil is shaken through a sieve. The soil particles can get through the holes in the sieve but the stones are trapped.

Filtering is similar to sieving, but it separates mixtures of liquids and tiny pieces of solids. Filter paper is paper with millions of microscopic holes in it. Liquids can get through the holes, but even very tiny pieces of solids are trapped. Another way of separating a solid from a liquid is to wait for them to separate. The solid particles will settle to the bottom of the container, leaving the clear liquid above. This process can take a few hours or even a few days.

· · · · · · · · · · · ·

This hiker's water purifier contains a filter that removes microorganisms from river water to make the water safe to drink.

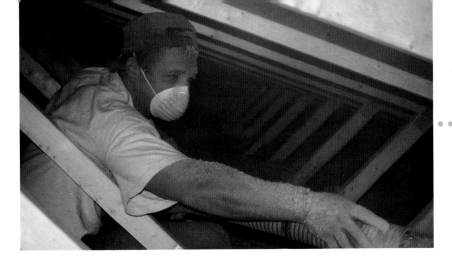

This construction worker wears a face mask when he works on a dusty site. How does the mask help protect him?

INVESTIGATION

How can you clean muddy water?

MATERIALS

Soil, some glass jars, filter paper, and a funnel.

INSTRUCTIONS

Pour some water into a jar. Add some soil to the water and swirl the water around to make a good mixture.

Fold a piece of filter paper into a funnel and rest the funnel in another jar. Pour the muddy water slowly into the paper, allowing the water time to drain through.

You will find that clean water pours into the jar, because the filter paper traps the soil particles.

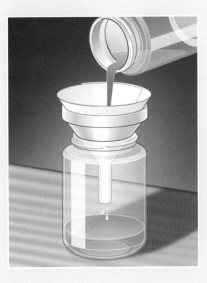

FURTHER INVESTIGATION

Oil and water separate into layers of liquid if they are mixed together in a container. If you add a detergent, what happens to the mixture? Can it now be separated out?

Filters are also used to clean air. How many examples of these filters can you find in your home, or in your family car? Try looking in stove tops and vacuum cleaners.

How can we separate solutions?

A solution is made when a material dissolves in a liquid. We sometimes need to separate out the solution. This is so that we can obtain the solvent without the solute or the solute without the solvent.

Distilled water is pure water, with nothing dissolved in it. It is often used by industries and in hospitals. But natural water has minerals such as salt dissolved in it. To make natural water into distilled, it has to be purified to remove the minerals. This is done by a process called *distillation*. The natural water is boiled in a container. The steam goes along a tube and is condensed back to water in a different container. The minerals are left behind, so the water in the second container is pure.

If there are two or more liquids that need to be separated, the same process can still be used. After the first liquid has boiled off, heating continues to boil off the second, the third, and so on. This process is used in industry to separate mixtures of liquids with different boiling points.

These salt flats are used to get salt from the sea. Sea water is put in shallow pools under the hot sunshine. What happens to the water?

INVESTIGATION

How do you get drinking water from the sea?

MATERIALS

A pan, a small cup, kitchen foil, salt, and ice.

INSTRUCTIONS

Pour a cup of water into a pan. To make drinking water like sea water, stir in salt until no more will dissolve (you will see the salt settled at the bottom of the pan). Put the empty cup in the middle of the pan.

Cover the pan with kitchen foil and press down the center of the foil slightly. Put some ice cubes on the foil.

Get an adult to help you gently heat the pan for a few minutes until the water boils. Be careful not to let it boil dry. Turn off the heat and allow everything to cool. Taste the water in the cup and in the pan.

Can you work out how the fresh water got into the cup? The water in the pan was very salty. What happened?

Crude oil is a mixture of different oils, which evaporate at different temperatures. Oil refining separates the crude oil by a process called *fractional distillation*. This picture shows a distillation tower at an oil refinery.

FURTHER INVESTIGATION

Can you find out what other dissolved substances are found in sea water? Can you find other uses of distillation in industry?

Glossary

Atmosphere
The blanket of air that surrounds the Earth.

Boiling
The change of state when a liquid turns into a gas. It happens at the substance's boiling point.

Change of state
When a substance changes from being in one state to being in another state. Boiling and freezing are examples of changes of state.

Compressed
When something is squeezed to make it take up less space.

Condensation
The change of state when a gas turns into a liquid. It happens at the substance's boiling point.

Conductor
A substance that allows heat or electricity to flow through it easily.

Crystal
A piece of solid material with a regular shape, such as a grain of sugar or salt.

Dissolve
To break up into tiny particles in a liquid. Sugar dissolves in hot water.

Freezing
The change of state when a liquid turns into a solid. It happens at the substance's freezing point.

Gas
One of the three states of matter. A gas flows and expands to fill the space it is in. Air is an example of a gas.

Heat
A form of energy. The more heat energy you put into an object, the higher its temperature gets.

Insulator
A substance that does not allow heat or electricity to flow through it easily.

Liquid
One of the three states of matter. A liquid flows to fill the bottom of the space it is in. Water is an example of a liquid.

Melting
The change of state when a solid turns into a liquid. It happens at the substance's melting point.

Mineral
A chemical that comes from the rocks of the Earth's crust.

Mixture
A substance made up of two or more solids, liquids, or gases mixed together.

Particle
A tiny piece of a substance.

Rust
A reddish-brown, flaky solid that forms on iron or steel that comes into contact with air and water.

Solid
One of the three states of matter. A solid does not flow. Wood is an example of a solid.

Solute
A substance that dissolves in a liquid to make a solution.

Solvent
The liquid that a substance dissolves in to form a solution.

State
Short for "state of matter." Either a solid, a liquid, or a gas.

Synthethic
A synthetic material is one that is not natural but made artificially.

Temperature
A measure of how hot something is.

Transparent
Used to describe a material that can be seen through clearly.

Volume
The amount of space that something takes up.

Water vapor
Water in gas form.

Further information

BOOKS

Changing Materials (Material World)
by Robert Snedden
(Heinemann Library, 2007)

Hands-on Science
by Sarah Angliss
(Kingfisher, 2002)

Materials (BBC Fact Finders)
by Martin Hollins
(BBC Books, 1996)

Materials and Matter (Making Sense of Science)
by Peter D. Riley
(Smart Apple Media, 2005)

Materials and Processes (Straightforward Science)
by Peter D. Riley
(Franklin Watts Ltd, 1998)

Materials and Their Properties (Heinemann Explore Science)
by Angela Royston
(Heinemann Library, 2003)

Materials Technology (Material World)
by Robert Snedden
(Heinemann Library, 2007)

Solids, Liquids, and Gases (Material World)
by Robert Snedden
(Heinemann Library, 2001)

CD-ROMS

Eyewitness Encyclopedia of Science
Global Software Publishing

I Love Science!
Global Software Publishing

ANSWERS

page 6 On the volcano, there is ice and steam. It is important to realize that clouds of steam are made up of tiny water droplets, so are not really water in gas form. There is a lake of water at the base of the volcano.

page 7 The carbonated drink contains all the states (the ice is solid, the drink is liquid, and the gas is carbon dioxide).

The use of liquids in machines is called *hydraulics*.

A road drill is worked by compressed air, which is a gas.

page 9 Only a diamond can be used to cut another diamond. The blade is covered in tiny crystals of diamond.

page 11 Metals are good conductors.

Graphite is not a metal but is a good conductor.

page 15 The movement of the particles and the way they are arranged changes, but the number of particles and the type of particles they are stays the same.

Ice cream with a high water content would melt at about 32°F (0°C), the melting point of ice. (It will not melt at exactly 32°F (0°C), because the water is not pure.)

page 17 Air expands when it is heated and contracts when cooled.

The balloon would expand slightly in the warm water.

Changing the pressure also makes the volume of a gas change.

page 19 Carbon dioxide was dissolved in the drink. When you open the bottle, the gas begins to come out of the solution, forming bubbles.

Solids dissolve more easily in hot liquids than cold ones.

Cooking oil is not a solvent.

page 21 Water evaporates from the sea, lakes, and land, and then condenses again to form tiny water droplets that make up the clouds.

page 23 The gas that blows up the balloon comes from the chemical reaction. The gas is actually called *carbon dioxide*.

The gas is made by a chemical reaction between two chemicals in the extinguisher.

page 24 Cooking causes a permanent change. The molecules in the food are affected by heat energy and there is a chemical change.

page 25 The gas that has been used up is oxygen.

To make a fire you need heat, oxygen, and fuel.

A candle changes from a solid to a liquid and finally to a gas before it will burn.

Some of the wax has been used up by the burning, but some liquid wax has run down the side of the candle and solidified again.

page 27 The face mask contains a filter that traps dust. It stops the worker from breathing in the dust.

The detergent causes a milky-white emulsion to form. This does not separate out.

page 28 The water evaporates.

page 29 The water boils, leaving the salt behind. Then it condenses under the cold ice and drips into the cup.

Sea water contains many dissolved substances, including calcium, sulfate, and carbonate.

As well as refining oil, distillation is used to make alcoholic spirits.

Index

Web Sites
Due to the changing nature of Internet links, PowerKids Press has developed an online list of Web sites related to the subject of this book. This site is regularly updated. Please use this link to access this list: www.powerkidslinks.com/sci/solid